The Best Seat In The Ho[...]

A User's Guide

Michael Edwards

Introduction 3
   Is a complete theory of consciousness possible? 9
   The theoretical and the practical 10
   A special case: dualism vs monism 10
   Religious dualism 12
Consciousness in the real world 13
   The interface between the thinking machine and the biological machine 13
      Thinking machine (TM): a working definition.................13
      The Neural Correlates of Consciousness.......................13
      The chemical correlates of consciousness: reward and punishment chemicals 14
   Consciousness creates access 16
   Conservation of consciousness 18
   Testing for consciousness 18
      Use of a polygraph test to test consciousness 19
Theoretical bases for consciousness 21
   Consciousness as a self-modelling construct....................21
   Meta thought..28
      Building a self-modelling thinking machine 29
      Testing for self-modelling consciousness.......................30
   Machine consciousness 31
   Thought experiments 33
      The Chinese Room 34
Conclusions 39
References 43

Introduction

Consciousness is the elephant in the room. It is how we see the world and how we interpret it, and how we see ourselves and interpret that too. In many ways, it is what we are, and it is our world: it is our very existence. It is, to paraphrase Fatboy Slim (Cook, Q, (1999)), Right Here, Right Now. It is the best seat in the house – indeed, it is the only seat in the house we are going to get, so we should try to appreciate it.

We perceive ourselves as individuals, with hopes, desires, fears, loves, hates, the whole gamut of emotions. We see ourselves as being free, as being able to make plans and act on them. We see ourselves as being moral agents. We act (we hope) morally and reasonably. All of this is through our consciousness, either phenomenologically through what we feel and see and hear and taste, or as conscious agents, thinking, planning, making decisions and acting on them.

But all of this is far from obvious, especially where this comes from. The religious among us may feel it comes from their god. But since the Enlightenment of the 17th century fewer and fewer of us believe in a god of this sort, even as mysticism creeps back into modern societies. An old philosophical principle, called non-overlapping magisteria, is sometimes invoked to say that science will never understand or have a correct theory of consciousness or the soul. This idea states that there are different realms of knowledge: in this case one of them being the physical, scientific world, and the other being the religious or spiritual world: the proposition concludes that it is impossible to understand a concept that exists in one magesteria with tools derived from another magisteria. In some ways it's a restating of Descartes dualistic mind body problem. Or of category mistakes (Ryle, G. (1949), Wikipedia contributors (2025)). The idea of non-overlapping magisteria is dismissed by almost all modern scientists and thinkers. It's not even clear if any of the major religions hold this view any more.

Consciousness cries out for an explanation. Long before humanity had explanations for gravity, evolution, light or much else, we had explanations for our own conscious existence, namely in the thousands of belief systems that sprang up in human societies. They served the purpose of a model, however inaccurate, of a cruel and arbitrary world in which we humans had no tools to investigate its true makeup. Indeed, were it not for the existence of consciousness, it seems incredibly unlikely that the concept of a god need ever have arisen. Firstly, there would be nothing to desire a god, and secondly there would be no consciousness (read spirit, soul etc) to explain.

Is it important to know what is conscious? Although we may not phrase it this way, or even think of it this way, most of us accept that a human being has rights based on their consciousness. On the basis that we are individuals, we have hopes and fears, we feel pain, we feel anxiety, we enjoy freedom, we hate captivity etc. Many of us attribute some or all of these rights to other animals. We don't attribute those same characteristics to a piece of metal or a lump of rock. Hence we don't feel much sadness in smashing up a rock or melting a piece of metal. If we don't understand fundamentally what the difference is between these different types of objects, the foundation for our morality in this area is on shaky ground. Many countries have legislation around human rights, indeed the United Nations has several charters in this area. The American Declaration of Independence, is in some ways a declaration of human consciousness and the rights that it confers. Many countries also have legislation around the treatment of animals, on the basis that they experience some, if not all of the same consciousness that we do. So there is a genuine, pragmatic reason to get as good an understanding as we can of consciousness, what creatures have it, do any machines have it and what rights (and, indeed, obligations) it confers.

Its understanding should also hugely help, not only in the treatment of mental conditions, but also in maximizing, in a semi-utilitarian sense, the happiness, or (less trivially), the self-actualization we can give to conscious creatures (or machines), on this planet or indeed elsewhere.

As to the title of this text (The Best Seat in the House), this terminology is this author's poor attempt at a philosophical joke. It's always a sign that a joke is poor when it has to be explained, so here is that explanation. Some early theories of consciousness had the idea that within the human head, perhaps within the brain, there was a little man (a "homunculus") that looked at the pictures coming in through the eyes, and the smells through the nostrils, and the sounds through the ears, etc etc, and reacted with pleasure or pain or any one of the many emotions, and then decided what actions, if any, he (always a he) wanted to take. In this way, he had "the best seat in the house". Indeed, the only one. The joke, such as it is, is that this theory of the homunculus is clearly garbage, and does not explain consciousness in the slightest. It begs the question, what is it inside the homunculus' head that does all the interpreting and action taking? A further, even smaller homunculus presumably. This leads to infinite regression and Turtle-ism (Wikipedia contributors (2025)).

Of course, any creator is almost the (in scale) physical opposite of a homunculus and simply adds an extra layer of complexity to any description of consciousness or, indeed, the world. As the old saying goes, it's pretty well turtles the whole way down (Wikipedia contributors (2025)). God is the reverse homunculus. It creates additional, possibly infinitely regressing complexity. Occam's razor (Wikipedia contributors (2025)) tells us to be wary of unnecessary complexity. As has been noted by wits: if God created man in his own image, we certainly returned the compliment.

Rationality and science seem to be the obvious place to look for an explanation for consciousness: after all science has explained many (if not yet all) of the things we see around us. The movement of the planets, the energy we receive from the Sun, the forces we experience of gravity, electricity and others. An understanding of human anatomy and physiology, along with pharmacokinetics and pharmacodynamics that have greatly enhanced human health and longevity. An explanation for the evolution of species of increasing complexity on this planet. And the inventions it has allowed, including amazing developments in computing, energy, communication and transport among many others.

But traditional science seems to offer very little hope of explaining consciousness. There's a great asymmetry in the manifestation of consciousness. Our experience of our own consciousness is often so all pervading that it does appear to be almost what we are, indeed, it may well be what we are, at least to ourselves. But that knowledge of ourselves is not reciprocated in the evidence we have for the consciousness even of other human beings, even those closest to us. Only by analogy can I look at another person, a friend, or even my wife and detect that they are conscious. Whereas potentially, you could measure my IQ, my height, my weight, the proportions of carbon, oxygen and nitrogen that I have in my body, the rate of firing of neurons in different parts of my brain, you have no way of measuring my consciousness. Indeed, there's no unit of consciousness unlike all those other properties just listed. This asymmetry is a huge problem for science. The scientific method is based on objective measurement and observation. And yet we can't objectively measure our own consciousness. Almost by definition, because we are both the subject and the object of the phenomenon. And when it comes to measuring the consciousness of another creature, another human being say, the problem is even more difficult. We rely on them reporting their feelings of consciousness to us. We use analogy. We try to infer consciousness through brain activity. And yet this is deeply flawed.

The literature of consciousness is peppered with references to "zombies". It's a questionable term, conjuring up images of artistically bereft horror movies of half dismembered rotting corpses blundering around a darkened film set trying to eat the brains of ordinary human beings. But its use in the consciousness literature is as a representation of, say, a human being, who looks and acts just like any one of us, but has no consciousness. Thought (and other) experiments can then be analysed into whether an automaton like this would act in the same way as a conscious person. There are serious, respected investigators in the field of consciousness who publish work denying the existence of consciousness (during the course of which they interestingly deny their own consciousness). Certainly, the existence of books on consciousness seems to indicate it is a real phenomenon. The only counterargument would appear to be parallel to Christians who deny evolution: God placed the fossils there, to test our faith or for some other reason. Perhaps there is a consciousness god or joker, who publishes books which expound its existence just to trick or test us. But really, if there is no real "us", who is the joker fooling? The sophistry seems overwhelming. See Wikipedia contributors. "World Turtle."

There is one further, third, possibility. That neither is everything material nor are there two different things for thinking things and physical things: everything is in fact, a thinking thing. This idea is called idealism (Guyer, P, Horstmann, R-P (2023)). For every single thing we see in the physical universe, at the core of each particle, like a tiny quantum (for want of a better word), is a non-divisible "particle" of consciousness. These particles of consciousness aggregate into larger objects such as ourselves in which the consciousness emerges as true agency, qualia, etc. The temptation of this theory is clear. For those of us who feel that the most real thing in the universe is our own feeling of self, of our own consciousness, and that everything else, even the physical world around us, is interpreted through this consciousness, why not argue that consciousness is the fundamental material? And hence the physical world is somehow instantiated out of consciousness too, not made fundamentally of quarks and bosons etc.

Hard to resist, maybe, but the world of science and engineering has produced incredible results and its predictive powers have been impressive. Based on principles and laws that govern such phenomena as gravity, the strong nuclear force, the weak nuclear force and electromagnetics etc it seems unlikely that underpinning all this is some sort of "quantum" of consciousness. Occam's razor, for one, seems to indicate that instantiating one further level of complexity for no obvious benefit is not a good plan. So this document won't take the idea of idealism any further.

This asymmetry, of what we know of our own consciousness but can only infer about other humans and other creatures, has led to some of the greatest problems in human history. The Nazi demonisation of other races as untermensch, barely human rats, when they're talking about Jewish people or gypsies is justifiable if you think they have a lower level of consciousness, a lower level of feeling, effectively a lower level of humanity. Of course, we know this is abject trash. But if you could actually prove that at least all humans have a similar level of consciousness (therefore a similar value and similar humanity) then this should lead to fewer wars based on demonisation and targeting of people who are different. Or at least the perpetrators might need to come up with new excuses.

There is a long and illustrious history to thought, discussion and science about consciousness. At least as far back as Socrates, Plato and Aristotle, and probably earlier. This book does not aim to cover the history of consciousness thought. Although it will reference some of the great thinkers of course. The Cambridge Handbook of Consciousness (version 2) (Zelazo, Moscovitch, Thompson (2023)), for those who want to get a thorough grounding in much of the history and the current state of the discipline, is highly recommended.

Lastly, it's possible that some non-scientific, religious or semi-mystical theory of consciousness, which may or may not include the idea of souls or spirits, is the correct one. However, in this short book

I've sought to follow the maxim that it is better to be wrong for the right reason than right for the wrong reason. By which it is meant that it is better to try to make progress on the question at the risk of being wrong, rather than retreat to myth and theology.

Is a complete theory of consciousness possible?

The human mind evolved to increase the chances of an individual surviving to sexual maturity, and perhaps a little longer to raise young to independence. It did not evolve to allow us to solve or understand everything in our universe. That includes consciousness. Our minds do not, we should assume, give us any direct insight into absolute reality. They are simply complex wetware machines that increase our chances of survival. If this means presenting a simplified, or even incorrect, interpretation of actual reality, that is what our minds will do. Hence, along with many other questions about the universe, both internal and external to us, we should not necessarily expect that we can come to a meaningful understanding of consciousness. That is not to say that we shouldn't try. The way we understand electricity, optics and gravity may not represent objective reality of those systems, but has undoubtedly led to benefit to mankind. In that science and engineering has been able to predict and hence, when applied, create devices and chemicals that we take so much for granted. So the pursuit of a credible theory of consciousness is, it is to be hoped, a worthwhile endeavour.

The theoretical and the practical

There are two main issues in focus here. The first is a valid theoretical model of how consciousness exists, how it works.

The second is how this is physically instantiated in our own universe. In the creatures and possibly machines that are, or could be, conscious.

Although the second question is without doubt interesting, the more interesting question to this writer is the former. It is clear from our own experience that the latter exists: after all we are thinking, conscious, moral agents (unless there's an awful lot of trickery going on somewhere). The substrates of this consciousness seem to vary quite a lot, with the likelihood that it exists in birds (with tiny avian brains quite different to our own), octopuses (with very different brains to our own), bats (who have no sight but who are very likely conscious) and possibly even simple nematodes that evolved about 500 million years ago.

Due to the different experiences we individually have of consciousness as human beings, along with the future (current?) possibility that machines or computers that we create are conscious too, the actual substrate seems less important: it's the rules and processes that allow consciousness, that support it and indeed create it, that are of greater interest. Not that this short work will completely ignore the second question, in large part as as the only thinking machines (this will be defined later) that we know are definitely conscious, i.e ourselves, are built on defined wetware, with brains made of neurons, and lots of supporting tissues and organs. Any discussion of consciousness would be foolish not to look at the concrete examples we have around us.

A special case: dualism vs monism

Dualism is too big a subject to be dealt with here, but it would be hard to have any kind of text on consciousness without a brief discussion of it. Rene Descartes' famous Cogito ergo sum ('*I think therefore I am*') has cast a very long shadow on the topic ever since he put those words down in the 17th century.

This book comes clearly from a materialistic view of consciousness: this is the prevailing view, but there remain some credible and respected scholars in the field who still rebut it. Perhaps the best known is the well-respected Australian philosopher and psychologist, David Chalmers. He uses the example of zombies (perhaps better described as flesh and blood automata) for a variety of thought experiments. That there could be human beings walking around planet earth exhibiting all the activities, actions and thought processes that we, conscious people, exhibit, but for whom there is no inner life, there is no agency, there is "nothing it is like" to be them.

In the Cambridge handbook of Consciousness (version 2) (Zelazo, Moscovitch, Thompson (2023)), mentioned previously, the example is given that it is possible to imagine a liquid with the appearance and all the apparent sensual properties of water. Something that can be drunk, is clear, tastes fresh, has no smell has no colour etc, and actually assuages thirst. But is not made up of two hydrogen atoms bonded to an oxygen atom, with trillions of them in a single drop of the liquid that we know is actually water. And so this is an analogy used with this automata argument. People do (or could) exist without consciousness who have all the appearances and all the characteristics of those with consciousness. In other words, consciousness exists outside what we physically see. To this writer, it appears there is a fundamental problem with this position. What would happen if you were to ask one of these automata if they were conscious? Even if they had read about consciousness, as they have no internal life (nothing it is like to be them, nothing it is like for them to taste honey, to smell a rose, to fall in love, etc), presumably they would have to answer that consciousness is an interesting proposition, like centaurs and time travel, but that it doesn't actually exist. The fact that many people can feel that consciousness exists (even though they may disagree in many ways about the form, nature and construction of it), surely indicates that this argument has a counter case, a black swan. When David Chalmers only sees white swans.

So an automaton asked the same question would behave differently to a conscious subject and reply that they were not conscious. Perhaps they would go on further to say that consciousness is a chimera, an illusion, a unicorn of the mind. So an automaton/zombie, that would appear identical to conscious humans is a concept that is imaginable, but like unicorns, cannot exist. To push it one stage further, perhaps in an attempt at a reductio ad absurdum, could an automaton/zombie ever write a book, without falsehood, propounding or accepting its own consciousness?

At base, this idea undermines the basic premise of the automaton / zombie argument. That an automaton would have different behaviour to a conscious subject in at least certain situations. The water-like liquid discussed above would reveal itself not to be water in a mass spectrometer, for example, if not to our senses.

Religious dualism

Perhaps the ultimate form of dualism is that espoused by some religions. Specifically, those where there is a spirit or a soul, generally with the nice side-effect of immortality. This is effectively how many religions view consciousness. But it's a very special sort of dualism. In many of them the dual entity, the res cogitans, is immortal. Also in quite a few, though this is perhaps a bit confused in some of them, there is no agency. Or at least, if there is, it's a very strange sort of agency. One in which there is predestination and predetermination (fate), which, prima facie, does appear to preclude agency.

Dualism in all its forms, property-based dualism (with perhaps Charmers' theories being the hardest to dismiss), along with Descartes' version, (substance-based dualism), won't detain us much longer in this text.

Consciousness in the real world

Despite stating earlier that this text was more interested in a theoretical model of consciousness, it would be nice to be able to detect it in the world in which we find ourselves, or at least some fingerprints or traces of it. Things that correlate with it or interface with it, or at least appear to do so.

The interface between the thinking machine and the biological machine

Thinking machine (TM): a working definition

This text uses a loose definition, but the term here can be taken to mean any machine, biologically based like the brains within people or octopuses or insects, or silicon based like modern computers, or other computational technologies, currently existing or hypothetical. These thinking machines can or do have capabilities to undertake analyses and actions that appear to be similar to biological thought processes in that they allow sophisticated interaction with the world in a purposeful manner.

The Neural Correlates of Consciousness

For most working in the scientific field of consciousness research (of who the majority are materialists), the understanding is that, for most life on earth that is conscious, the substrate for that is the brain.

With increasingly sophisticated equipment now available to look at brain activity, the link between consciousness and brain activity can be more easily studied (though still in a limited way). Indeed, some attribute consciousness studies gaining respectability in the late 1980's and early 1990's with the introduction and refinement of this new technology, with growing interest to the present day (eg Tononi G, Koch C (2008)). Specifically it allows, with far greater sophistication, assigning certain conscious states to certain neural activity, the so-called, neural correlates of consciousness (NCC).

This is a deeply specialised subject area that this writer is nowhere near qualified to comment on it in any depth. University College London's MetaLab Consciousness Club (http://metacoglab.org/consciousness-club, @ConsciousnessClubUCL) is a highly recommended source of videos of lectures on current investigation in this (and other) areas of consciousness research, given by their own scientists and global investigators.

Save to say that this area of science is making rapid strides in correlating brain activity with conscious and non-conscious states, and in so doing is providing greater evidence for the materialist view of consciousness.

The chemical correlates of consciousness: reward and punishment chemicals

Reward chemicals like dopamine and serotonin, and the natural endorphins and morphine-like chemicals that are released after, for example, orgasm, would appear to be unnecessary if there were nothing (no individual) to reward. It's possible to understand why fight or flight chemicals get released, to prepare the organism to help it survive. It could be said that the reward chemicals released after orgasm are to encourage further orgasms, therefore the ongoing and prolific multiplication of the species. The big difference here is that their benefits are to be felt in the future. Not right now.

But we see organisms that we are pretty sure are not conscious, (bacteria for example) that have no or small nervous systems and still reproduce hugely without consciousness. These rewards simply are not required and neither is punishment.

The reward (or pleasure) chemicals released when satiety is reached we can certainly register. There is a pleasant feeling to being full. Similarly, if not more intensely, to having had an orgasm. But there has to be *something that feels that satisfaction*, that pleasure, that reward. It would appear that any creature that has similar releases of these reward neurotransmitters, dopamine in particular, in similar situations to human beings, is also being rewarded, and therefore the supposition is fairly that it is also conscious.

Like pleasure or reward, producing pain is pointless, and it adds an unnecessary cost if there's no conscious entity to experience it, if "there is no one home". Pain chemicals are pointless for an unconscious creature. It's easy to understand how steroids, naturally produced within the body, white blood cells, increased blood flow, inflammatory action etc is all very useful for the physical healing of a creature. One that has been injured or perhaps has an infection or disease. But pain is pointless to this acute healing process. It can only act in the future. So, if one of the points of chemical punishment, perhaps the only point, is to encourage the creature to avoid the action that caused the painful situation, there has to be "a something" to encourage. For an unconscious creature pain would have no benefit whatsoever, and, as mentioned earlier, just has an overhead of a useless chemical being produced at some cost with no benefit.

So studies that look at the production of punishment and reward chemicals and the neurological and cellular receptors for them, presumably would be quite rewarding in determining what creatures are probably conscious.

The question also arises of how did these complex reward or pain chemicals arise evolutionarily?

It is possible to see that the production of reward and pain stimuli and chemicals, which is so dependent on a consciousness to feel those rewards and pain, is highly beneficial evolutionarily, in terms of making the individual attempt to gain reward, and attempt to avoid pain. It is also possible to see that any organism that could instantiate those mechanisms would survive far better than one that didn't; i.e. this quite possibly is the reason why consciousness is so desirable evolutionarily. Even though phrasing it like that makes it sound like consciousness is somehow teleological, which it clearly isn't. It is just another in the long line of tiny beneficial mutations amongst the far more common harmful ones, that have grown to give the possessor an advantage.

Even in simple and ancient creatures like Cnidarians and nematodes, (organisms like the roundworm C. elegans) there is dopaminergic neurotransmission. In the roundworm, it modulates locomotor activity, simple food-seeking behaviours, and even a basic form of reward-based learning. Although dopamine exists in simpler (evolutionarily older) organisms, it has no receptors, so cannot have been used for reward. Potentially the chemical was co-opted as a neurotransmitter as it could perform that function and was already present for other purposes in some evolutionarily earlier species. The evolutionary hurdle to reusing is lower than building from scratch.

We really should consider that even some relatively simple organisms that have a neural and chemical basis for consciousness, i.e. a brain of some minimal complexity, and dopamine receptors and some other neurotransmitters that would allow the transmission of pleasure and pain, might be conscious.

From Google's AI engine:
"The simplest organism known to use dopamine in a mechanism related to reward-seeking and learning is the nematode worm Caenorhabditis elegans.
In this very simple animal, which possesses a nervous system of only 302 neurons, dopamine is released from mechanosensory neurons in the presence of food (bacteria). This dopamine release modulates the worm's motor behaviour, causing it to reduce crawling speed and engage in an "area-restricted search" pattern of locomotion, which helps it to stay in an area with abundant food".

This nematode and its close relatives evolved perhaps 500 million years ago. They are so simple they have no respiratory or circulatory system. Therefore, it is possible that reward mechanisms were in place in these simple organisms as long ago as 500 million years. It suddenly brings into sharp focus the question of whether we should consider cows, or bees, or modern-day worms (basically a lot of the life forms around us), to be conscious or not. The supposition probably should be that they are conscious, and deserve a degree of our respect and care, rather than not, until we can prove otherwise.

Consciousness creates access

Consciousness creates access to the unconscious (McGovern, K. Baars, B. J.(2023)) Experiments in human subjects have shown that, in many cases, once they are made conscious through feedback of a previously unconscious process, the subject is able to control it consciously. This is often called biofeedback. The term has become controversial and although there are many practitioners using it ethically, some charlatans have used the genuine effect to sell treatments of varying sorts for varying conditions that have little or no effect. This should not detract us from accepting the underlying phenomenon which has been proven in studies published in peer-reviewed journals.

One example of biofeedback is that of controlling alpha waves in the brain. These occur at 8 – 12 Hz in frequency, and are believed to show the brain in relaxed wakefulness and rest. A person trained with biofeedback can take take conscious control and alter the magnitude of these waves. Spinal motor unit control is another example. Spinal motor units are a combination of muscle and nerves, that are normally under autonomic (non-conscious) control. They can be controlled consciously after biofeedback training. So if concurrent conscious input is given, for example sound through a set of headphones when electrical activity is detected in a unit, the subject, after very limited training (thirty minutes or so) can gain access to the activation of these motor units, even when the biofeedback is subsequently removed.

There are other examples of being able to elevate a previously subconscious function to consciousness, such that consciousness then has access to it. For things like typical and common conscious functions, like controlling finger movements during typing, we take this for granted. It is the ability of consciousness to be able to be trained to control other, previously hidden, actions that is so remarkable. Hence the substantiated claim that consciousness creates access to the subconscious.

I will quote an entire passage from Chapter 8 of The Cambridge Handbook of Consciousness as it seems so apt:

Consciousness seems to be needed to access at least four great bodies of unconscious knowledge: the lexicon of natural language, autobiographical memory, the automatic routines that control actions, and even the detailed firing of neurons and neuronal populations, as shown in biofeedback training. Consciousness seems to create access to vast unconscious domains of expert knowledge and skill (McGovern, K. Baars, B. J.(2023))

Conservation of consciousness

In the physical world several fundamental properties are governed by conservation laws. Conservation of energy and conservation of momentum for example. Sub-atomic particles themselves may not be conserved in collisions or nuclear fusion or fission, but the way they break down is governed by rules. Charge, spin, and energy are always conserved. Conservation of chemical elements applies in a predictable form too: elements have a half-life and slowly decay to other elements. That again is predictable.

If the materialist view of consciousness is correct, is consciousness also a property that is conserved? That there is some form of immortality, an immortal spirit or soul to use religious terminology? After all, if consciousness is some property, emergent or otherwise, of the physical universe, wouldn't that make sense? However, it's hard to see emergent properties being conserved when the substrates they emerge from disappear or decay. If society is an emergent property of groups of people, when the groups of people are taken away society disappears too. Any thinking machine, conscious or otherwise, does need to run on some hardware. Whether that is tin and silicon and copper, or complex hydrocarbons built into neural networks, a substrate is required. Until there is an ability to transfer a consciousness running on one system to another system, we cannot hope for longevity beyond the lifespans of our physical bodies. It's possible, however, to see that a consciousness based on constructed computational hardware, like the computers we are so familiar with, might be transferable from one host to another and hence be conserved.

Testing for consciousness

The Turing Test (Turing, A (1950)), invented by the famous mathematician and logistician, is now pretty well defunct. The idea that, if an averagely intelligent adult human being couldn't tell whether they were communicating with another human being or some other entity, based on a conversation and questions and answers etc, then the other entity must be conscious, is now seen to be fairly facile. It appears that, even if it looks like a duck and quacks like a duck, it may not be a duck. Mainly because we have not looked at enough "ducky" characteristics to be certain. We also need to look at how it walks, its wings, and its behaviours, etc, etc. The Turing test was always too simple to be a good test for consciousness, though perhaps that is a lot easier to say with the benefit of hindsight. Many of us, I suspect, have interacted with voice prompts on the phone when we contact large businesses. These are getting more and more sophisticated, to the extent that only slight nuances in accents, cadence, and lack of variability let us know that this isn't an actual human being we're communicating with. The example of the DeepMind AlphaGo (BBC, (2016)) beating the global Go world champion four to one in a series in the mid 2010's, and appearing to show creativity as part of its strategy, is a small but remarkable example of this. We no longer just have machines where their logic can be relatively clearly seen as a human implemented algorithm. Machine learning is getting to the point where it looks a lot like creativity and inventiveness, characteristics commonly associated with agency and consciousness. Of course, this will likely only increase as we go forwards, though with the Turing test now an effective irrelevance, it's hard to know how we would ever determine, due to the basic asymmetry of the experience of consciousness, whether a machine of this sort actually is sentient. Certainly it's hard to believe these Large Language Models (LLMs) are sentient currently, but that may just be an innate superiority complex of wetware-based thinkers over hardware-based ones.

Use of a polygraph test to test consciousness

The polygraph, or lie detector test, is not allowed in court in most countries, the United States being the obvious exception. However, it seems recognised that the polygraph test can indicate lying or lack of truthfulness even if most countries do not think this is to an acceptable standard to be admitted in a court of law.

So, what would happen if a non-conscious subject were asked the question "Are you conscious?" whilst being given a polygraph test? A polygraph test, simply put, looks for increased mental and physical activity, such as brain stimulation in certain areas, increased sweat, increased heart rate and other vital signs, as possible or probable symptoms caused by the subject lying. Consequently, if a subject were not conscious and were asked this question, they have two possibilities. They could answer no (the truth) that they were not conscious or they could answer yes (a lie). Would the polygraph test pick up the difference? Of course, if every human being is conscious, (which is certainly very possible), the answer "Yes" for a person is a truthful one. But it would be interesting to see if more subjects failed this test than normal false negatives that occur with a polygraph. There have also been more experimental, though successful, tests using FMRI as a way of detecting falsehoods. Langleben DD, Hakun JG, Seelig D, Wang AL et al (2016).

As to testing hardware-based thinking machines, perhaps ones based on the Clarion (Sun (2016)) model or any other, is it possible to conceive of a lie detector for a hardware thinking machine? Perhaps CPU cycles would parallel brain activity; heat production or data IO be an analogue of sweating? Is it possible to foresee setting up the rules and preconditions for asking an advanced non-biological thinking machine if it is conscious, and using some measures of this sort to indicate answers to two independent but related questions:

1. Is the TM conscious?
and/or
2. Is the TM capable of lying?

If the TM answers yes to the first question, do we take it at its word? If we don't, and we believe the machine is lying, then to lie you need to have a theory of mind, or a theory of knowledge (the terms seem to be used interchangeably). It is not enough to tell a falsehood, as that might be done in all innocence. The TM in this case would need to know that what it is propounding is inaccurate. It also needs to understand that whatever it is communicating with has a mind, or possible knowledge. This theory of knowledge is effectively meta knowledge: it is knowledge that there is knowledge. As such, if the TM is able to lie (however we determine that) it has probably come some way to proving it has much of the mental machinery necessary to be conscious

Theoretical bases for consciousness

Consciousness as a self-modelling construct

It seems an a priori truth that we plan based on where we see ourselves in the future: evaluating among a set of possibilities and actions and where they would likely leave us in the future: generally speaking, aiming for a better position than we are in now. Less hungry, fitter, having a comfortable temperature, more likely to find a mate, more likely to impress a potential mate, etc. We "see ourselves" in our plans. This is not necessary for agency, nor survival of course. A smart (but not conscious) thinking machine (TM) can evaluate the current situation, a favourable end point (without feeling any "desire"), and do a gap analysis. For example, the stomach of this unconscious creature is sending neurological or chemical signals that it is empty. Blood sugar level monitoring systems indicate low sugar. There is a calorific, nutritious plant nearby. The creature/machine formulates a plan to move towards that plant and consume it. Some minutes later, stomach distension has improved as noted by nerve feedback, and blood sugar levels are starting to recover. That plan and the same actions may well have occurred to a conscious machine, and, from the outside, there need be no difference in what the objective observer sees. But from the inside, the conscious creature has the sensation of what it is like to eat the calorific plant but the unconscious, not. And in the automaton, although a plan was formulated, it was formulated to satisfy overcoming some granular, non-optimal conditions that the TM was experiencing. In the conscious subject that self-models and plans based on that model, is there any qualitative or quantitative advantage to having this self-model?

Models are a highly efficient way of encapsulating a complex object, concept or scenario within a label, that means each time that model is encountered (or one close to it), the TM is not starting from the ground up and wasting valuable mental time and effort assembling granular sensory inputs, correlating them with granular memories for its "what if" analysis. They are a very useful shorthand. We can start our instantiation of a model with the basics: a "Tree Model" might start out in a thought process with just that label: our intelligence and senses identify that an object is very likely "a tree". If necessary, we might flesh out some basic characteristics: big, static, not a predator, big linear brown bits, small granular green bits that together cover a large area, and, functionally, the whole providing potential shelter. Sometimes occurring in groups in a "copse", "wood" or "forest". All information we know and store away (however that works) in the neurons that do "tree-ey" stuff for us. A more useful example is maybe that of Predator. To survive, it's critical that we identify a predator quickly. We've filed away in memory a set of (variable but related) characteristics, in a "Predator Model". We can instantiate that when our senses pick up enough information (sharp teeth, fast moving, large) for the mind to say: "Hang on, that looks damn like this Predator Model I have". What do we do when we have a high likelihood of one of those present? The model tells us to run, hide or fight, depending on other factors, like our understanding of other models: are "Helper" models present, can I outrun this particular predator, are there any "Hiding"-like structures close by? As we can see, these two factors are also usefully encapsulated into models.

The type of modelling being described here is often also referred to as Representationalism. Our perception of the universe is not an objective, exact picture of reality, but an internal representation of it (see arguments elsewhere in this text). What we actually perceive, what our brains represent to us, it's not objective reality, but what evolution has found to be useful for us, with our limited mental machinery.

The symptoms of low energy availability (what a conscious creature might describe as "hunger") could also be satisfied by consuming a prey animal. Let's model that in the TM:

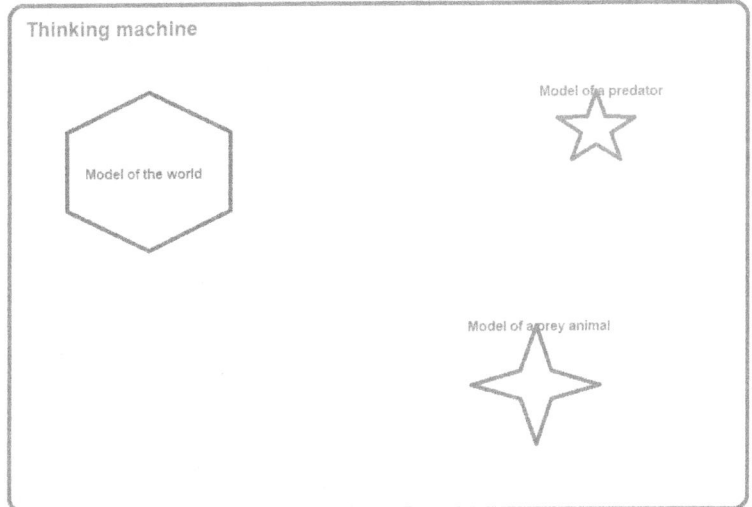

Note that this diagram of a Thinking Machine has plenty of capacity (empty space) for computation and analysis of things other than the world, a predator and a prey animal. As it must have – there's plenty more to staying alive and reproducing than finding prey and avoiding predators.

Note that the Model of the World does not imply that the TM has an anything like a complete model of the world, just that it has a working model of some important global characteristics of the world. Like scale (large and small), orientation (up and down, left and right) , hot and cold, fast and slow etc. There is certainly no implication that the TM understands the world in anything approaching its entirety.

Note also, that the encapsulation of a set of neurological conditions such as described above, into a label called hunger, or a colour called red, (what David Chalmers might call qualia), is a highly economical way of dealing with these far more complex underlying phenomena. Indeed, this efficiency might explain qualia along with other models. Qualia are simply labels for complex sensory input.

As to the deeper nature of qualia, that is for another section. However, how is the mental apparatus supposed to raise a complex set of stimuli into consciousness? As we've evolved from pre-linguistic creatures, they couldn't possibly use language within the brain to encapsulate a relatively complex (or even simple) sensation. Consider running your hand across the grain of a piece of wood. There is a feeling to this, there is something it is like to run your hand across a piece of natural but cut wood. There is possibly a faint smell of the resin remaining. Conceivably, with linguistic brains, we could have machinery that raises this to consciousness with the words "cut wood". But how would a pre-linguistic creature, or a human being without language, raise it into consciousness? It needs to be encapsulated in some other unique way. To my mind, that is simply what qualia are, a pre-linguistic form of encapsulating a sensation or a complex set of sensory inputs, into something the brain will interpret quickly. It could be argued, in some ways, that this is a statement of Higher order Thought Theory (Wikipedia contributors (2025)), with simple unconscious thoughts about grain and texture, smell, warmth and planarity, giving rise to a high order thought that encapsulates all of these. This helps bring qualia into the broader realm of consciousness in general.

We have evolved to reward ourselves when we identify a complex concept, such as hunger or predator, or up, or down, etc ad infinitum, with a small dopamine/serotonin release and a feeling of what it is like to detect a predator, to see the colour red, to taste a banana. In this model, qualia are explained (at lease partially) as the reward mechanisms for our brain doing a good job. And what point is there to this evolutionarily beneficial reward, if there is nothing to be rewarded, no agent, no sentience? Reward chemicals by themselves indicate a level of consciousness. Addiction appears to be a pathological extreme desire for rewards, requiring larger and larger amounts.

So what if a different (arguably more advanced) TM has a model of itself?

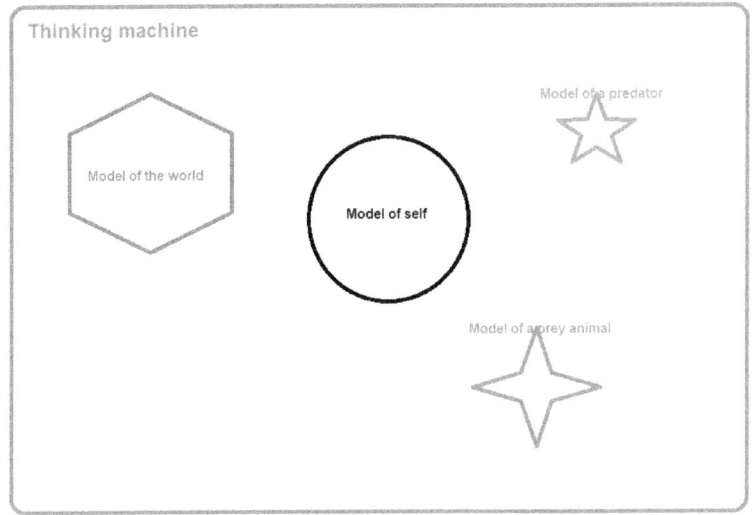

It seems obvious if the TM is able to model external concepts, it is only a small jump to modelling the organism itself. It also seems a self-evident truth that, as long as the overhead of modelling the self is not too high, and does not detract from the other chores the TM needs to do, that a self-model is likely quite useful. After all, the world does have a "me" in it: my world model can only be more complete, and hence more useful, if that is included in the modelled world. This does not, ex nihilo, mean that this model is necessarily seen as qualitatively different to all the other models by the host machine. However, it will probably be used in almost all planning exercises, unlike most other models. It will also have a great deal of customisation available at the will of the host: this object is under the control of the host as it is the host's representation of itself.

I am proposing that this self-modelling, this reflexivity, is what consciousness is. This is a far from original thought, and is sometimes labelled the self-model theory of subjectivity (SMT) (Metzinger T. (2007))

But does a problem arise if the TM tries to make a complete model of itself? Could we find ourselves with all capacity taken over by this, as the complexity of the entire self must be at least as complex as the TM, which is, after all, only a part of the self.

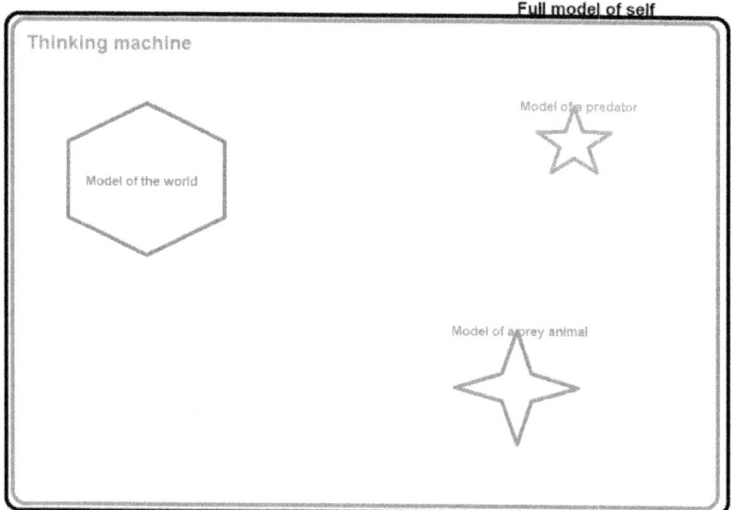

Here the outer boundary of the self-model has completely filled the TM, and in fact needs some more space which the TM cannot provide. Indeed, perhaps this is better shown like this:

Full model of self

It is not possible for a TM to completely model itself. An entity of complexity C cannot model another (or the same) entity of complexity C (or C + D), as the modelling activity itself must take some resources from the TM. It also ceases to have any capacity for anything else. So the modelling of the self must be partial, perhaps very partial.

If this self-modelling of consciousness that is being proposed cannot fully model itself, does it mean that consciousness is a con or a fragment of the "real me"? It certainly appears that it is only partial. But maybe at any one time a conscious entity can model in toto certain elements of itself that are pertinent to the situation. After all, we don't really need everything, everywhere all at once very often, if at all. And our experience of consciousness varies. Sometimes we are alert to much more in our consciousness. Sometimes, perhaps for reasons of reducing energy consumption, for rest and restoration, our physiological and mental systems decide to reduce the effort put into consciousness. Maybe our intellectual abilities are being put into other activities, unrelated to consciousness, the filing of memories, the reorganization of current structures and memories within the mind, and many others. So the fact that our consciousness may vary and will always be incomplete at any one point in time, is what we experience and is what is to be expected.

There will be those who will say, with some legitimacy, that the feeling of being conscious does not feel in any way like a self-modelling experience. Or that it is far more than that. To which the reply might be given that the feeling (qualia) of eating a banana, is really nothing like a banana. That the smell of a rose is unrelated to the three hundred or so volatile chemicals in the scent (Chromatography Today (2025)). Consequently, we should not be surprised if the extensive catalogue of qualia that make up consciousness do not appear directly related to any theoretical model of the same.

Meta thought

In the Cambridge Handbook of Consciousness (Zelazo, P. D, Moscovitch, M and Thompson, E (2023)), Drew McDermott, page 136, writing on AI, introduces, effectively, a thought experiment. Imagine a chess-playing robot. It's able to move, see chess boards, and engage in simple conversations such as "Would you like to play chess?". McDermott builds up his scenario. It's possible to see that a hard-wired robot might recognize standard chess pieces and the standard chess board if well lit, with the correct perspective, the correct scale etc. But McDermott gets us to consider perceptual errors. So the chess pieces might be a bit bigger or a bit smaller than expected, or not the usual ebony and ivory colours, or perhaps not the usual shapes. As we know, there are many different sorts of novelty chess sets or indeed elaborate sets. As a human being, we could probably recognize a Star Wars based chess set, or, in his example, one based on Disney characters. For the robot, he gets us to consider that it might have modules for improving its task performance. Such that feedback and insight into its own cognitive/computational activities allows it to fine tune its interpretations of input and hence its actions. Initially it may not recognize a Disney character chess set, but feedback of whatever form allows it to improve its computational/cognitive abilities to recognise a chess set and chess board of various types. Potentially, also to identify different kinds of chess players, perhaps being able to determine that, although no obvious human adversary is present, the chess set itself is a chess computer, and that again satisfies its requirement for an opponent. At this stage the machine is having two levels of thought: the simpler, task related computation about identifying a chess set and a chess board and a potential opponent, and then the second level, meta-thought, or the thought about the task related thoughts it's having. Thus enabling it to make better decisions by constantly reviewing its task related thoughts, their level of success, and fine tuning them for future use. This is almost a form of what is discussed elsewhere in this text, of having a self-model, although in this case it's a very limited self-model, merely of its own computations about chess equipment and players, and improving its ability to complete tasks in that arena. It is, nonetheless, a kind of meta thought.

In the BBC series, Secrets of the Brain (BBC, 2025), narrated and presented by Jim Al-Khalili, early in the second episode we're shown actual experiments and we see a mouse with an electrode implanted in its brain, negotiating a simple maze with the Predator, a robot that puffs air at it, pursuing it. The electrode brain sensor is able to show that, before the mouse moves in a certain direction, its brain is already considering this move. In the programme, this is described as imagination, which is probably a good term, but we could equally use the term consciousness. The mouse is imagining itself in the future, at a different location, and is considering whether that location will be likely closer to food and further away from the robotic predator. Imagination could be considered another form of meta thought.

Building a self-modelling thinking machine

It should be possible, at least in theory, to produce a computer program that self-models. Maybe this has already been done, eg Connectionist Learning with Adaptive Rule Induction On-line (CLARION, Sun, (2016)). In some ways, a self-modelling program would be an iterative program, ie able to iterate itself. And possibly could be placed in a world, (perhaps an open world similar to that in some video games), to experience and navigate by itself.

Computational and scientific literature has many articles on software of this sort (eg Landauer, C., Bellman, K.L. (2003)). Some of it describes the systems as "self-aware" without specifically using the term "conscious". It's an interesting discussion as to whether a system that is self-aware is, by definition, conscious. I propose that it is: though it may not have agency, that is not necessary for consciousness.

Testing for self-modelling consciousness

Can we envisage some tests to help establish whether, firstly, a self-modelling mechanism is occurring, and secondly, if it is, is it giving rise to consciousness?

The writer is no expert on psychological and neurological investigation, clinical or laboratory, but poses the question: would the internal representation of a self have any external correlates, anything that an investigator could see or detect? It may be possible to use introspection and self-reporting, but these are complex ideas that our subject could quite likely misinterpret, and the flaws with introspection as a form of investigation of mental states and consciousness are well known.

It seems possible, at least theoretically, that some humans are not conscious. Indeed, in the last few years a number of studies, (eg Hurlburt, R.T., Alderson-Day, B., Kuhn, S. & Fernyhough, C. (2016)), found that between 20 and 30% of people report having no internal monologue / dialogue (anendophasia). (Note, I refer to the internal dialogue or internal monologue interchangeably. The "conversation" is clearly within one individual, hence the term monologue might seem more appropriate, but as much of it is a set of questions and answers or comments on statements, the actual feeling of it is more of a dialogue).

Aphantasia, the inability to conjure up mental representations of things, eg. a unicorn or a memory of a beautiful beach, also exists in a substantial proportion of people when questioned. For many of us, both of these are an important part of our consciousness. Clearly, however, we cannot and should not discount consciousness in people with anendophasia or aphantasia and they may be as fully conscious as everyone else. But these are interesting findings.

Some of the pathological activities undertaken by sociopaths and psychopaths (for example) seem more likely to occur in people, if such exist, without consciousness. It appears that consciousness, (and having a self-model with which one applies, by analogy and similarity/empathy, consciousness to other people or animals), means such a person is less likely to undertake pathological or sociopathic activities against others. Activities such as violence, manipulation, rape and murder. But that's only a postulate. As a conscious person the writer knows that he has committed activities that could easily be described as antisocial or sociopathic and he has felt regret. Regret is another emotion that requires a consciousness.

Perhaps a polygraph could be used here too. Although the test would have to be skilfully constructed and the questions very well posed. Performing this on anendophasics and aphantasics would also be interesting. One hopes and expects the answer yes, to the question "Are you conscious?" and that the polygraph would indicate they were not lying. But it would nonetheless be an interesting study to carry out, assuming ethics panel approval.

Machine consciousness

At the current stage of the understanding of consciousness, and in the attempt to build conscious machines, one practical or empirical problem that faces thinking machines convincing humans they are conscious, is that we know they were built by humans. Unless, that is, we hide them in boxes, behind non-natural interfaces like keyboards and screens, as is done in the Turing Test (Turing, A (1950)). Without that obfuscation, we will know that they have routines running inside them, however complex, that were human designed and built. That they are built on silicon (possibly some other substrates in future) and use an instruction set or programming language or languages, all created by human beings. This makes it harder for us to make the leap that they could ever be conscious. After all, our experience with any other machine, including sophisticated computers, indicates that they are not conscious.

With a dolphin, or an elephant, an octopus or a chimpanzee, the workings of the brain and the mind inside these creatures is still very much a mystery, however far we've come in the neurological and cognitive sciences. This mystery allows us more easily to attribute another mystery, 'consciousness', to them. Attributing consciousness to something that, at least theoretically, we know everything about: its construction, its processes, its source of energy, its memory units, its processors etc etc is no doubt difficult. Superficially, a modern computer needn't look much different to a toaster, and we're not likely to attribute consciousness to that (although maybe, in a few years time, we might have to). It's no wonder that we would find it hard to think of a 'mysterious' property like consciousness emerging from such mundane and lifeless materials. And that's part of the problem. Although something like the Turing Test hides the actual substrate of the possibly conscious machine from the interrogator, in the end we know that we will recognize a robot or a human being as one or the other. The world of *Blade Runner* and almost human-indistinguishable replicants has not yet arrived, but perhaps isn't that far away. But we're not there yet. In the end, the substrates for biological consciousnesses are pretty simple too: carbon, hydrogen, oxygen, cells, mitochondria, neural connections, sodium ions, potassium ions, electrons: basically a whole lot of organic wetware. But the actual fact that we don't manufacture these ourselves seems to allow us to supervene the characteristic of consciousness onto some of the organisms with these wetware components, which we find much harder to do with a man-made machine.

There may also be an element of modesty. The idea that, were we to create such a conscious machine, we would be (hubristically) playing God. Yet the mainstream of consciousness thinking accepts that early life was not conscious. The machinery had to get more complex. So what that there wasn't a designer or some teleological force, there was evolutionary force. If, as this writer believes, any thinking machine that has models of itself (along with its environment); attempts prediction; uses abstraction, aggregation, modelling and symbols for representing complex artefacts of its world; and can plan and retrieve episodic events from memory to assist in its planning for the future, is likely to be conscious: whether it's made of metal and silicon, or, alternatively, biological molecules.

The ancient and celebrated game of Go (Wikipedia contributors (2025)) has been subject in recent years to the creation of machine competitors. Most famously the AlphaGo application (BBC News, 2016) in the 2010's. This application ended up finishing world Go champion four games to one against a human grand master. At times it was noted that it would take much longer than at others to make a single move. There could be simple computational reasons why this occurred. And in this case that's perhaps the most likely cause. DeepMind, the makers of AlphaGo, also monitored other aspects of the activity of the application and hardware as it played the game. Consider the (currently hypothetical) situation of a thinking machine designed to be conscious, perhaps based on a self-modelling concept. A polygraph test asking the question "Are you conscious?" could be monitored to see if the level of activity was higher, or approximately equal to, that predicted if the machine believed the statement to be true. If the thinking machine, the computer application in this case, came to the answer rather directly, it may be concluded that it saw the assertion as an a-priori truth, requiring few CPU cycles and round trips to its memory. Or whether, in fact, a lot of computation and time was involved in an affirmative response, in which case one might be more likely to suspect the machine was "lying".

This is a test that would require a lot of planning and a lot of thought to be run fairly and to give a true indication on whether the machine was conscious or not. Even then, the results would be open to interpretation and dispute.

The Clarion model (Sun, Ron (2016)) is a computer system designed to emulate one particular possible model of consciousness. It has been a very useful model in looking at machine consciousness, and consciousness in general. As to whether the machine is actually conscious, it's not clear whether the creators believe it is or not. Perhaps we will never know if androids dream of electric sheep. See Dick, P K (1968).

Thought experiments

The Chinese Room
John Searle (Searle, J. (1980)), the American philosopher, came up with a very interesting thought experiment which he published in 1980. It has caused much discussion and controversy ever since.

Essentially, and somewhat simplistically, he was trying to use his experiment to prove that Strong AI (the ability for a simple computer with hardware and software to be conscious), is logically impossible.

He defined Strong AI as the belief that, "The appropriately programmed computer with the right inputs and outputs would thereby have a mind in exactly the same sense human beings have minds." This kind of computer is sometimes called a Turing machine.

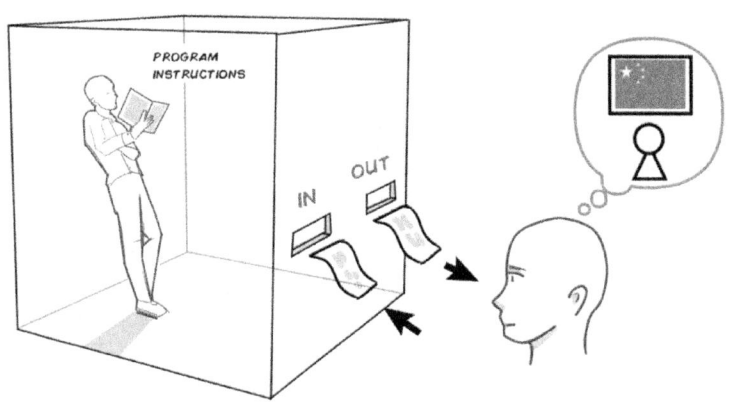

The experiment itself consists of a conceptual physical situation. In a large room sits someone who understands Chinese. In a smaller room within that room, sits a person, perhaps Searle himself, who does not understand Chinese. A small hole in the inner room allows the Chinese speaker in the outer room to pass written messages into the inner room. The person (who does not understand Chinese) inside that room has a rule book that allows them to convert the incoming text into outgoing Chinese text of a relevant response. That person then passes their formulated (but not understood) reply out through the second hole back to the Chinese speaker.

Simple versions of the experiment can be run and the outputs can convince the person in the outer room (the Chinese speaker) that there is a Chinese speaker, or at least an entity that understands Chinese, in the inner room. As this is a thought experiment, it isn't generally actually run, but it can be imagined that a sufficiently comprehensive and accurate rule book would allow the non-Chinese speaker to produce appropriate responses without having any understanding, or developing any understanding, of Chinese. The conclusion to be drawn, Searle would suggest, is that where any machine appears to show cognition and understanding (and therefore possibly consciousness), this is a trick. Much like the person inside the inner room, no understanding has taken place.

There are big faults with the Chinese Room thought experiment as a means of proving (or attempting to prove) that the computation model of consciousness is incorrect, indeed impossible. There is the full systems view refutation. This proposes that, although the non-Chinese speaker does not understand Chinese, the *system*, of the person plus rule book, does, which seems to be a strong counter argument. After all, individual parts of the brain of a Chinese speaker cannot be said to understand Chinese, but the entire brain undoubtedly does.

A more important flaw with Searle's hypothesis is the fact that the Chinese Room only works with the simplest of questions and answers and not with a series of related questions: a conversation. One of the main reasons for this is that the Chinese Room is stateless. As such, there has to be no relationship between each question and answer pair, because the non-Chinese speaker inside the inner room has not altered in his or her understanding of Chinese. He or she has also not gained any useful reference memories from the previous interaction. They remain a blank slate. It's also impossible to meaningfully answer a follow-up question, if you don't even know it's a follow-up question and don't recall the previous question, nor know what it was about! So this person remains in the same, ignorant, state as at the start of the entire experiment. This does not change as each question is received and each answer generated. Purely in the terms of participation in the experiment, the person in the inner room does not change state so each question is taken as a stand-alone.

So if, for example, the human Chinese speaker outside the room asks the question "Have you ever been to China?" the inner room might reply "No". If the Chinese questioner then asks "Why?", as the room itself is stateless, the question "Why?" has no context. It is not, and cannot be, interpreted as relating to the previous question and answer. Therefore, how can the inner room give a meaningful answer? "Why" *what*, precisely? Why is the sun shining? Why did France and Germany go to war in the 1870s? Why does light travel at exactly the speed it does? This lack of statefulness is so different from that of a human being's mind, that the experiment becomes almost meaningless. It's also very different to the way a digital computer or Turing machine would work as well. Digital computers hold state these days, and have done for a very long time. They know, for example, that when you pay some money into your bank account, the state of the account prior was that you already had some credit balance and so the new amount is added to the previous total. Simply put, any computer for the last 50 years (and certainly the human brain/mind) has memory and is able to hold a context of its current situation in numerous ways. Indeed, that is a critical part of consciousness.

Searle could argue that he didn't exclude memory and statefulness from his thought experiment, even though it's not explicitly stated and even though most demonstrations of his thought experiment don't include memory/state. But imagine the complexity of the reference materials, the rule book used by the human being inside the room, if state were to play a part. Not only would the incoming symbols, (the syntax), have to be matched to a reference book and the output found, but every single previous input would also have to be part of the matching. As the question "Why?" or "Why not?" might refer to just the previous reply, or a conversation that goes back an hour, a day, or even weeks, (with multiple questions and answers in between) it is clear that the problem rapidly escalates. Searle might say this is just a practical difficulty, but it's certainly one hell of a practical difficulty. Indeed there could even be questions from the Chinese speaker of the sort "What did you think about Mao's Great Leap Forward?". In what way does the Chinese room respond, when an opinion, not a fact, is being requested? And yet we see modern large language models, (which, of course Searle would not have had experience of when he came up with this thought experiment in 1980), are able to come up with answers to such open-ended questions.

Indeed, Searle might suggest that I'm being far too literal in my exposition of his thought experiment, that the rule book could in fact be a computer that had memory and algorithms within it: but were he to, he would simply adding an extra level of complexity and it would no longer the Chinese room experiment as most of us understand it. The experiment would be open to the homunculus argument. He's proposing a Turing machine inside a Turing machine, and after that its pretty well turtles the whole way down.

Searle's thought experiment fails in almost the exact opposite way to Turing's. Turing takes what is now seen as the over simplistic view that, if a machine responds in a way that convinces a human that it is conscious, it must be conscious. This falls down if one accepts that large language models are already able to impersonate humans in a simple task-oriented conversation, and yet very few people think they are actually conscious. Searle's experiment, on the other hand, fails to produce a machine that can jump the low bar of appearing conversationally human, let alone conscious. The lack of context, memory and follow-up in responses would clearly immediately allow a person to see that the Chinese Room is not a conscious entity.

One clear problem is that, in the experiment, the person in the room is standing in for the hardware of the computer, with the book of rules being the software. Therefore, to claim that the hardware has no semantics, no comprehension in this case, is to be expected: it is the software, the rules engine, that has the 'understanding'. Hence why the full systems view is such a convincing refutation. It would be like saying that the brain is the conscious entity in higher animals, whereas almost everyone agrees that it is what is *running* in the brain, the 'software': the complex patterns of interaction of neural excitation (the neural correlates of consciousness) or (in the case of this thought experiment), the book of rules, that, if anything in the room, might be conscious.

At the very best, the Chinese Room thought experiment shows that an incredibly limited form of digital computer is not able to be conscious. It does not show that complex computational machines, thinking machines, stateful machines, could not be conscious.

Conclusions

I hope this brief book has outlined some of the wonder of consciousness, if you did not feel that already. It is my hope that it has shown that the apparent magic that it is, is actually explicable, if not now, probably in the near future. And this explanation will be given in clear, physical and procedural ways, with recourse to nothing beyond the known physical laws, even if some of these laws are not fully developed yet. There is certainly no reason to think the world is made of two things, as Descartes thought: the physical world and all the forces that act in that, and a mental or thinking world, made up of something quite different and non-overlapping. And, lastly, it certainly does not need belief in the supernatural, in religion or a creator.

The most convincing theory of consciousness to this writer, is that any system capable of complex self-modelling in time and space will be conscious. Obvious strategies for dealing with the complexity of the world, such as top-down prediction rather than exclusively bottom up interpretation; abstraction, encapsulation and labelling; reward and punishment neurotransmitters that reinforce beneficial behaviour all support this functionality.

These capabilities enable the multiple characteristics and sensory inputs that constitute a probable predator to be encapsulated into a concept or label of 'predator' which allows that mental representation to be manipulated, and for its actions to be predicted quickly. It doesn't take much to see that if that abstraction and encapsulation not only includes external objects and concepts in the world, but also the self, planning around potential future needs and risks to the self is going to be highly optimised for survival and eventual passing of genetic material on to offspring, hence reinforcing these characteristics.

There are many similarities between self-modelling and modelling the external world. Both rely on memory, encapsulation, evaluation, and planning. The sensory input for the external world (extraception) is the ability to sense external information and interpret it effectively. Conversely, the sensory input for the self, is introspection. This introspection probably explains why the internal monologue/dialogue, is such an important part of being a conscious entity (despite some studies, mentioned earlier, that indicate this may not be the case for everyone).

How accurate does this self-model have to be? Given that the model humans, (and therefore presumably to some degree, thinking creatures and machines), have of the external world is fairly limited and probably inaccurate, and given that biological systems purely create these models to survive and procreate, not as any teleological aim to a genuinely close approximation to actual reality, if the self-model the thinking machine has is partial or inaccurate in certain respects, that is no worse than the modelling it has of the external world.

The main thing the self-model has to have that is different to the other models, is the concept of agency. That, unlike the models of the external world which can only be indirectly affected, this internal model is, at this level, the thinking machine, and can propose to act in almost any way that the machine has the capability to instantiate. A good self-model will rapidly filter out any proposed actions that it could take that are not within its capabilities. A self-model that did not fulfil this would be deviant, and evolution would soon filter it out. As an example, if we, as reasonably feeble biological thinking machines, had a self-model that imagined we could pacify a sabre-toothed tiger and carry it back to our tribe as a source of sustenance, it would lead to catastrophic consequences for the individual and hence the swift removal of any genotypes that sustained this deviant modelling from the gene pool.

So, simply put, the self-model need only be practical, and separate itself from the other (external) models by the attribute of agency, for it to be useful.

This self-model then quickly leads the thinking machine to almost continuous hypothesising about the various possibilities for the self-model in amongst the external models, and what actions could be taken to most benefit itself. This constant hypothesising is the internal monologue or dialogue that most of us experience and that's probably the most intense, or at least pervasive, symptom of consciousness.

The substrate of that may be the grey wetware we have in our own brains and nervous systems, or those (fairly similar) in primates, some mammals, and (less similarly) birds. To those much less similar: possibly many insects, fish, and even invertebrates such as nematodes, whose close relatives evolved around 500 million years ago. This does seem to imply that if we affect these organisms, perhaps hurt or kill them, we are doing wrong. Their rights should be taken into account in our decision making.

It does mean that machine consciousness is also possible and may even already exist on planet Earth. And, as with our wetware brothers and sisters, our hardware cousins will need respect, if and when they acquire consciousness.

The existence of reward and punishment chemicals appears to be good evidence for consciousness. You cannot punish a rock nor reward it, as there is no inner life, no "I" or "you" to reward or punish in order to change future harmful behaviour or reinforce current beneficial behaviour. In this way, we can see that consciousness provides a huge evolutionary advantage, giving the creature clear temporal markers of what will give it the greatest utilitarian reward and the least punishment, that parallels reproductive and evolutionary success. This leads to an obvious conclusion: if an organism has these neurotransmitters, they should be considered conscious and worthy of our care.

The self-model that is proposed here and has been proposed by far more qualified people (eg Landauer, C, Bellman, K.L. (2003)), perhaps not uniquely amongst theories of consciousness, does fit in nicely with the existence of reward and punishment chemicals. The self-model must exist over time. The reinforcement of good behaviour by reward chemicals and the avoidance of bad behaviour by the reinforcement of punishment chemicals is a clear chemical way of embedding good behaviours into the planning and hypothesizing of the individual. Reward and punishment are not concepts that could exist within a thinking machine with no self-model. There would be nothing to reward or punish.

Although we may not feel that our magical and mysterious consciousness, the 'here and now'-ness of our existence, to be much like a self-model, neither is the glorious smell of a fragrant rose much like the 300 chemicals that make up that perfume. But the latter is definitely true, and so the apparent disconnect cannot be used to dismiss the veracity of the former.

We should treasure our brief time with these marvellous thinking machines that evolution has provided us with and hosted us in, and be as kind as we possibly can to other creatures or systems that are, or probably are, or even just may be, conscious.

References

Cook, Q, *Right Here, Right Now*, Skint, 19 Apr. 1999

Ryle, G. *The Concept of Mind.* University of Chicago Press, 1949

Wikipedia contributors. *Category mistake.* Wikipedia, The Free Encyclopedia. 15 Jun. 2025, retrieved 8 Nov. 2025.

Wikipedia contributors. *World Turtle.* Wikipedia, The Free Encyclopedia, retrieved 2 Oct. 2025

Wikipedia contributors. *Occam's razor*, Wikipedia, The Free Encyclopedia. Retrieved 7 Oct. 2025.

Guyer, P, Horstmann, R-P, *Idealism, The Stanford Encyclopedia of Philosophy (Spring 2023 Edition)*, Edward N. Zalta & Uri Nodelman (eds.).

Zelazo, P. D, Moscovitch, M and Thompson, E, *The Cambridge Handbook of Consciousness*, Cambridge University Press, 2023

McGovern, K, Baars, BJ. *Cognitive Theories of Consciousness, The Cambridge Handbook of Consciousness, Chapter 8,* Cambridge University Press, 2023

Turing, A. *Computing Machinery and Intelligence*, Mind, 1950

Langleben DD, Hakun JG, Seelig D, Wang AL, Ruparel K, Bilker WB, et al. *Polygraphy and functional magnetic resonance imaging in lie detection: a controlled blind comparison using the concealed information test*. J Clin Psychiatry. 2016 Oct;77(10):1372–80

Dick, PK, *Do Androids Dream of Electric Sheep?*, Doubleday, 1968

Artificial intelligence: Google's AlphaGo beats Go master Lee Sedol 2016, BBC News.

Secrets of the Brain, BBC2 Television (2025)

Sun, Ron (2016). *Anatomy of the Mind: Exploring Psychological Mechanisms and Processes with the Clarion Cognitive Architecture.*

Oxford University Press.

Landauer, C, Bellman, K.L. (2003). Self-modelling Systems. In: Laddaga, R., Shrobe, H., Robertson, P. (eds) *Self-Adaptive Software: Applications. IWSAS 2001.* Lecture Notes in Computer Science, vol 2614. Springer, Berlin, Heidelberg. https://doi.org/10.1007/3-540-36554-0_18

Wikipedia contributors. *"Higher-order theories of consciousness."* Wikipedia, The Free Encyclopedia. 20 Jun. 2025. Web. 7 Nov. 2025.

Metzinger T. (2007). *Empirical perspectives from the self-model theory of subjectivity: A brief summary with examples.* Models of Brain and Mind - Physical, Computational and Psychological Approaches.

Wikipedia contributors. *Go (game).* Wikipedia, The Free Encyclopedia. 14 Nov. 2025, retrieved 29 Nov. 2025.

Hurlburt, R.T., Alderson-Day, B., Kuhn, S. & Fernyhough, C. (2016). *Exploring the ecological validity of thinking on demand: Neural correlates of elicited vs. spontaneously occurring inner speech.* PLoS One, 11(2), e0147932.

Chromatography Today (2025), *What Do Roses Smell Of? — Chromatography Investigates.* https://www.chromatographytoday.com/news/gc-mdgc/32/breaking-news/what-do-roses-smell-of-mdash-chromatography-investigates/46607 (accessed 7 November 2025)

Searle, J. *Minds, Brains, and Programs.* Behavioural and Brain Sciences. 1980

Tononi G, Koch C. *The neural correlates of consciousness: an update.* Ann N Y Acad Sci. 2008 Mar;1124:239-61. doi: 10.1196/annals.1440.004. Erratum in: Ann N Y Acad Sci. 2011 Apr;1225:200. PMID: 18400934.

Printed in Dunstable, United Kingdom